Name_____ Date_____

If you want to book a virtual math lesson by the creator, you can email stelaniemarshall@gmail.com. Do not hesitate If you have questions, or if any teachers have any topics (especially common core related), they would like added to the workbooks contact me. If I decide to add your subject manner, I will mention you and your school in the workbook.

Some later pages are better in groups or may need teachers/ parents to help. I made some pages and problems easier while others closer to the next grade level. I like to add work that will take time and really need for the child to focus.

Name_____ Date_____

Name_____ Date_____

Dragon Lesedi's Magical Magnificent Bakery
2nd grade Book 1
Stelanie Marshall LLC

Name_____ Date_____

Name_____ Date_____

Name_____ Date_____

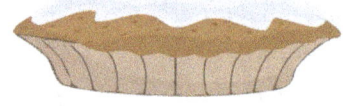 A Blender of Review

1) 425
   +133

2) 7+_____=15

3) 30+40+10=_____

4) 6+4+5=_____

5) 777
   -432

6) 50-20=_____

7) 99-27-41=_____

8) 20-13-3=_____

<>or=

9) 7_____13

10) 15_____15

11) 7+3_____5+4

12) 19-6_____17-3

pattern

13) 10 _____ 30 40 _____ 60 _____ 80 90 _____

14) _____ 40 60 _____ _____

15) 5 _____ _____ 20 25 30 _____ 40 45 50 55 _____ 65 _____

_____ 80 _____ 90 _____ 100

Count money totals:

16) 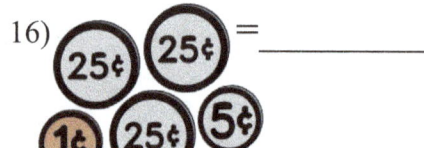 =_____

17) 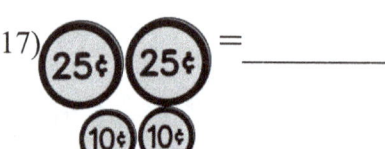 =_____

18)  =_____

Dragon Lesedi's Magical Magnificent Bakery 2nd grade 1 Stelanie Marshall LLC

Name_____ Date_____

Work Area/ Chefs Corner

Name_____ Date_____

Restocking Regrouping Addition

Sample
$$
\begin{array}{r}
9\\
78\!\!\!/6\\
+401\\
\hline
1190
\end{array}
$$

Sample
$$
\begin{array}{r}
4\\
3\!\!\!/68\\
+260\\
\hline
628
\end{array}
$$

1)
$$
\begin{array}{r}
477\\
+171\\
\hline
\end{array}
$$

2)
$$
\begin{array}{r}
195\\
+732\\
\hline
\end{array}
$$

3)
$$
\begin{array}{r}
369\\
+524\\
\hline
\end{array}
$$

4)
$$
\begin{array}{r}
716\\
+637\\
\hline
\end{array}
$$

5)
$$
\begin{array}{r}
347\\
+144\\
\hline
\end{array}
$$

6)
$$
\begin{array}{r}
528\\
+315\\
\hline
\end{array}
$$

7)
$$
\begin{array}{r}
235\\
+680\\
\hline
\end{array}
$$

8)
$$
\begin{array}{r}
239\\
+913\\
\hline
\end{array}
$$

9)
$$
\begin{array}{r}
68\\
+791\\
\hline
\end{array}
$$

10)
$$
\begin{array}{r}
539\\
+455\\
\hline
\end{array}
$$

11)
$$
\begin{array}{r}
108\\
+\ \ 34\\
\hline
\end{array}
$$

12)
$$
\begin{array}{r}
831\\
+539\\
\hline
\end{array}
$$

Name_____ Date_____

Work Area/ Chefs Corner

Name_____ Date_____

 Greater Than, Less Than or Equal to.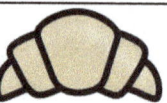

1) 67 _____ 76   2) 80 _____ 79   3) 48 _____ 51

4) 4+4 _____ 10-3   5) 12+4 _____ 20-3   6) 19-1 _____ 14+5

7) 789 _____ 643   8) 456 _____ 546   9) 71 _____ 74

10) 108 _____ 111   11) 601 _____ 610   12) 389 _____ 398

13) 9 _____ 27   14) 15 _____ 25   15) 57 _____ 54

16) 52-31 _____ 60-40   17) 75-62 _____ 35-20

18) 100-40 _____ 30+40   19) 100-20 _____ 32+43

Dragon Lesedi's Magical Magnificent Bakery 2nd grade 1 Stelanie Marshall LLC

Name_____ Date_____

 Balance the orders

1)15+4=_____+10   2)5+8=2+_____   3)9-_____=18-13

4)7+4=_____+10   5)19-6=_____-4   6)7+3+4=_____+10

7)15+10+5=20+_____   8)5+1+17=15+_____   9)13+14=48-_____

Circle the correct numbers the correct color for each group.

10)Even- 8   3   41   7   9   22   76   110   85   134

11)Divisible by 5 – 43   100   78   65   20   57   23   15   18   5

12)Odd- 7   48   77   52   94   109   29   66   51   139

13)Divisible by 10- 35   80   12   59   170   93   140   60   190

14) One of the flower warlocks finished a powerful remedy before going to grab some pancakes. She used 15 tubes of red snail glitter, 12 tubes of green hopping leaves and 10 tubes of purple midnight howls. How many tubes were used in all?

_____

Name_____ Date_____

Hundreds, Tens and Ones.

391 _____ _____ _____
        Hundreds   Tens   Ones

107_____ _____ _____
        Hundreds   Tens   Ones

545_____ _____ _____
        Hundreds   Tens   Ones

934_____ _____ _____
        Hundreds   Tens   Ones

729_____ _____ _____
        Hundreds   Tens   Ones

863_____ _____ _____
        Hundreds   Tens   Ones

432_____ _____ _____
        Hundreds   Tens   Ones

Name_____ Date_____

Hundreds, Tens and Ones.

956_____ _____ _____
     Hundreds   Tens   Ones

345_____ _____ _____
     Hundreds   Tens   Ones

718_____ _____ _____
     Hundreds   Tens   Ones

239_____ _____ _____
     Hundreds   Tens   Ones

867_____ _____ _____
     Hundreds   Tens   Ones

180_____ _____ _____
     Hundreds   Tens   Ones

421_____ _____ _____
     Hundreds   Tens   Ones

Name_____ Date_____

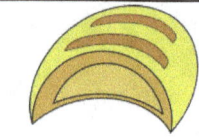 Thousands, Hundreds, Tens and Ones.

7985_____ _____ _____ _____
　　　Thousands　　Hundreds　　　Tens　　　　Ones

3471_____ _____ _____ _____
　　　Thousands　　Hundreds　　　Tens　　　　Ones

5349_____ _____ _____ _____
　　　Thousands　　Hundreds　　　Tens　　　　Ones

1237_____ _____ _____ _____
　　　Thousands　　Hundreds　　　Tens　　　　Ones

4603_____ _____ _____ _____
　　　Thousands　　Hundreds　　　Tens　　　　Ones

2758_____ _____ _____ _____
　　　Thousands　　Hundreds　　　Tens　　　　Ones

7195_____ _____ _____ _____
　　　Thousands　　Hundreds　　　Tens　　　　Ones

6512_____ _____ _____ _____
　　　Thousands　　Hundreds　　　Tens　　　　Ones

Name_____ Date_____

Thousands, Hundreds, Tens and Ones.

5332_____ _____ _____ _____
     Thousands   Hundreds   Tens   Ones

1891_____ _____ _____ _____
     Thousands   Hundreds   Tens   Ones

7139_____ _____ _____ _____
     Thousands   Hundreds   Tens   Ones

6574_____ _____ _____ _____
     Thousands   Hundreds   Tens   Ones

3297_____ _____ _____ _____
     Thousands   Hundreds   Tens   Ones

4703_____ _____ _____ _____
     Thousands   Hundreds   Tens   Ones

9415_____ _____ _____ _____
     Thousands   Hundreds   Tens   Ones

8926_____ _____ _____ _____
     Thousands   Hundreds   Tens   Ones

Name_____ Date_____

Who's Who Ticket Number? Color in the correct square.

	605	559	836	520

Clues

1-The knight's diner ticket starts with a 5.

2- The queen's ticket is an even number.

3-The dragon's is divisible by 10

Name_____ Date_____

Who's Who Ticket Number? Color in the correct square.

	294	288	265	379
 (sheep)				
 (green giant)				
 (pirate giraffe)				
 (yellow knight)				

Clues

1- The yellow knight's ticket is an odd number

2- The Pirate's ticket divisible by 5

3- The sheep's number ends in an 8

Name_____ Date_____

Regrouping With Subtraction.

732
-481

645
-136

432
-217

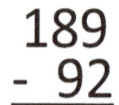

189
- 92

334
-227

912
-413

879
-199

449
-384

276
- 83

619
-481

512
-341

298
- 99

Dragon Lesedi's Magical Magnificent Bakery 2nd grade 1 Stelanie Marshall LLC

Name_____ Date_____

Work Area/ Chefs Corner

Name_____ Date_____

Greater Than, Less Than or Equal to.

603_____630   583_____385   720_____711

466_____426   809_____908   208_____208

25+8+5_____19+14   34+12+7_____29+26

138+77_____118+98   22+66_____43+34

673-434_____555-311   421+279_____200+500

700-500_____623-418   7+9+3_____10+4+5

1+7+9_____5+5+8   25-10-5_____30-15

179_____197   280_____268   351_____335

Name_____ Date_____

Match the Dates so Lesedi gets the desserts out in time.

1/2/2xxx   November 6

2/1/2xxx   December 7

11/6/2xxx   July 12

10/7/2xxx   January 2

12/7/2xxx   June 11

6/11/2xxx   February 1

7/12/2xxx   October 7

Name_____ Date_____

Mini Stories at the Bakery and Diner. When more than 1 add the number that many times.
Practice for multiplication tables.

Pancake Specials
The Castle's Stack- 10
Jester's Mayhem-5
Magic Wand- 8
Warlock's Buttery Potion- 7
Kingdom's Welcome-9
Unicorn Leap-4
Genie Wishes-3
Pirate's Chest-6
Double Moons-2
Single Moon-1

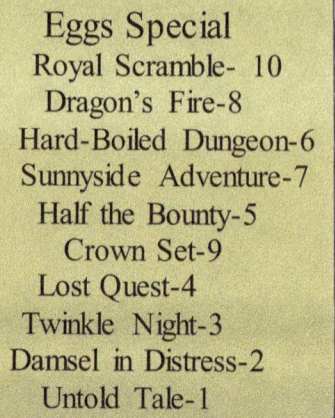

Eggs Special
Royal Scramble- 10
Dragon's Fire-8
Hard-Boiled Dungeon-6
Sunnyside Adventure-7
Half the Bounty-5
Crown Set-9
Lost Quest-4
Twinkle Night-3
Damsel in Distress-2
Untold Tale-1

Example- An order of 3 Unicorn Leaps is 4+4+4=12=3x4

1-Ser Jamal was famished and decided to make a large order and asked for 2 Pirate's Chest and 3 Half the Bounty's in his order.

How many were ordered? Eggs_____ Pancakes _____

2-One of the snow fairies needed to pick up 5 orders of Magic Wand and 6 Dragon's Fire.

How many were ordered? Eggs_____ Pancakes _____

3-**(Take your time and do one at a time if needed) Tim the Giant came in after finishing up at his jewelry store. He ordered 10 orders of The Castle's Stack and 8 orders of Sunny Side Adventure.

How many were ordered? Eggs_____ Pancakes _____

4-A red knight came in and ordered 3 Warlock's Buttery Potions and 4 Crown Sets.

How many were ordered? Eggs_____ Pancakes _____

Name_____ Date_____

Work Area/ Chefs Corner

Name_____ Date_____

5-Carl the dragon came and ordered 8 Magic Wands and 5 Royal Scrambles after he helped the blacksmith start a fire.

How many were ordered? Eggs_____ Pancakes _____

6-Princess Saman from the north came buy while passing through to drop off new paperwork for the royal committee and picked up 2 Genie Wishes and 4 Lost Quests.

How many were ordered? Eggs_____ Pancakes _____

7-Malcolm the genie ordered 8 Single Moons and 3 Damsel in Distress.

How many were ordered? Eggs_____ Pancakes _____

8-April the juggler juggled her order of 5 Kingdom's Welcomes and 8 Twinkle Nights out of the diner.

How many were ordered? Eggs_____ Pancakes _____

9-Linzy and Mali came in early in the morning and picked up 9 Jesters Mayhems and 11 Sunnyside Adventures.

How many were ordered? Eggs_____ Pancakes _____

10-Risa the fairy godmother asked for extra syrup with 5 Unicorn Leaps and 7 Dragon's Fire.

How many were ordered? Eggs_____ Pancakes _____

11-Timothy a crystal apple farmer grabbed 6 Warlock's Buttery Potions and 3 Royal Scrambles.

How many were ordered? Eggs_____ Pancakes _____

12-Paris nibbled on 3 Double Moons and 8 Hard Boiled Dungeons.

How many were ordered? Eggs_____ Pancakes _____

Name_____ Date_____

Work Area/ Chefs Corner

Name_____ Date_____

Intro to Multiplication.

This plate represents 3x4 and 4x3

4+4+4=12

3+3+3+3=12

Lesedi had to prepare for a huge cake for a gargoyle wedding, so she got 4 cartons of 6 eggs=6x4.

6+6+6+6=24 eggs in all. Also 4+4+4+4+4+4=24.

Name_____ Date_____

Adding Up the Orders with Multiplication.

1) 4x5=_____

2) 2x6=_____

3)4x7=_____

4)5x5=_____

5)8x8=_____

6) 1x1=_____

Name_____ Date_____

Adding Up the Orders with Multiplication.

6x11=_____

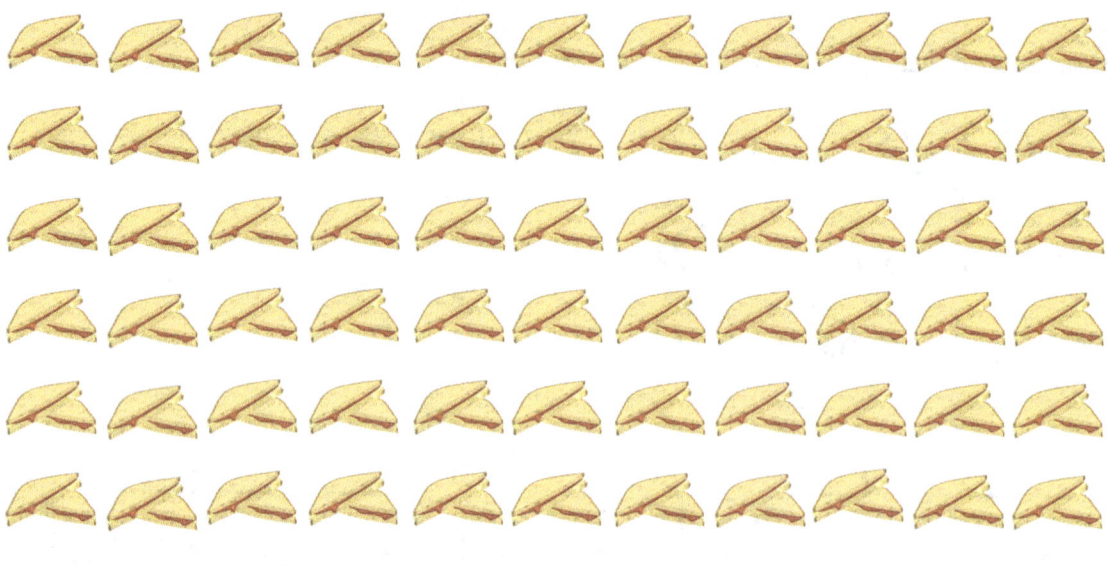

9x3=_____

2x2=_____   4x4=_____   2x3=_____

Name_____ Date_____

Adding Up the Orders with Multiplication.

6x3=_____

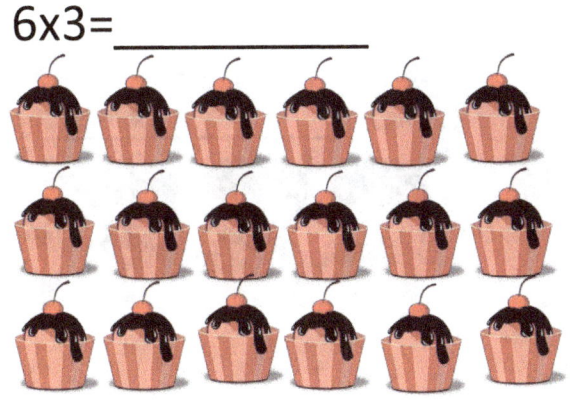

5x7=_____

12x2=_____

10x3=_____

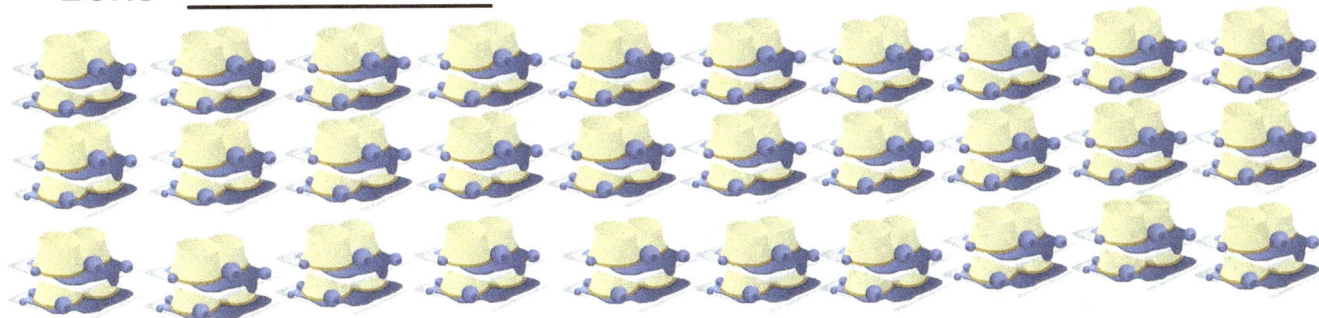

8x3=_____

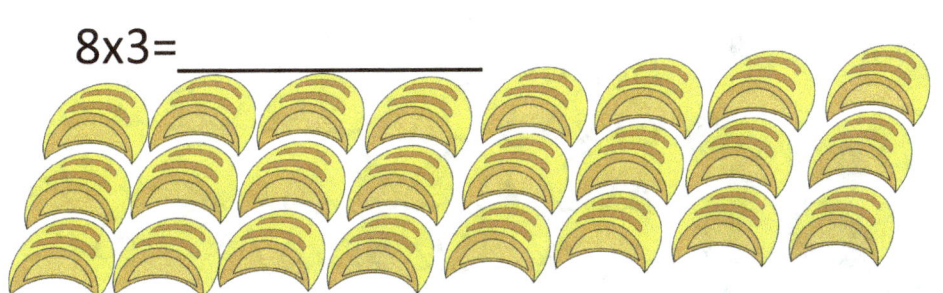

Name_____ Date_____

Adding Up the Orders with Multiplication.

6x1=_____

2x3=_____

3x3=_____

1x9=_____

7x6=_____

5x4=_____

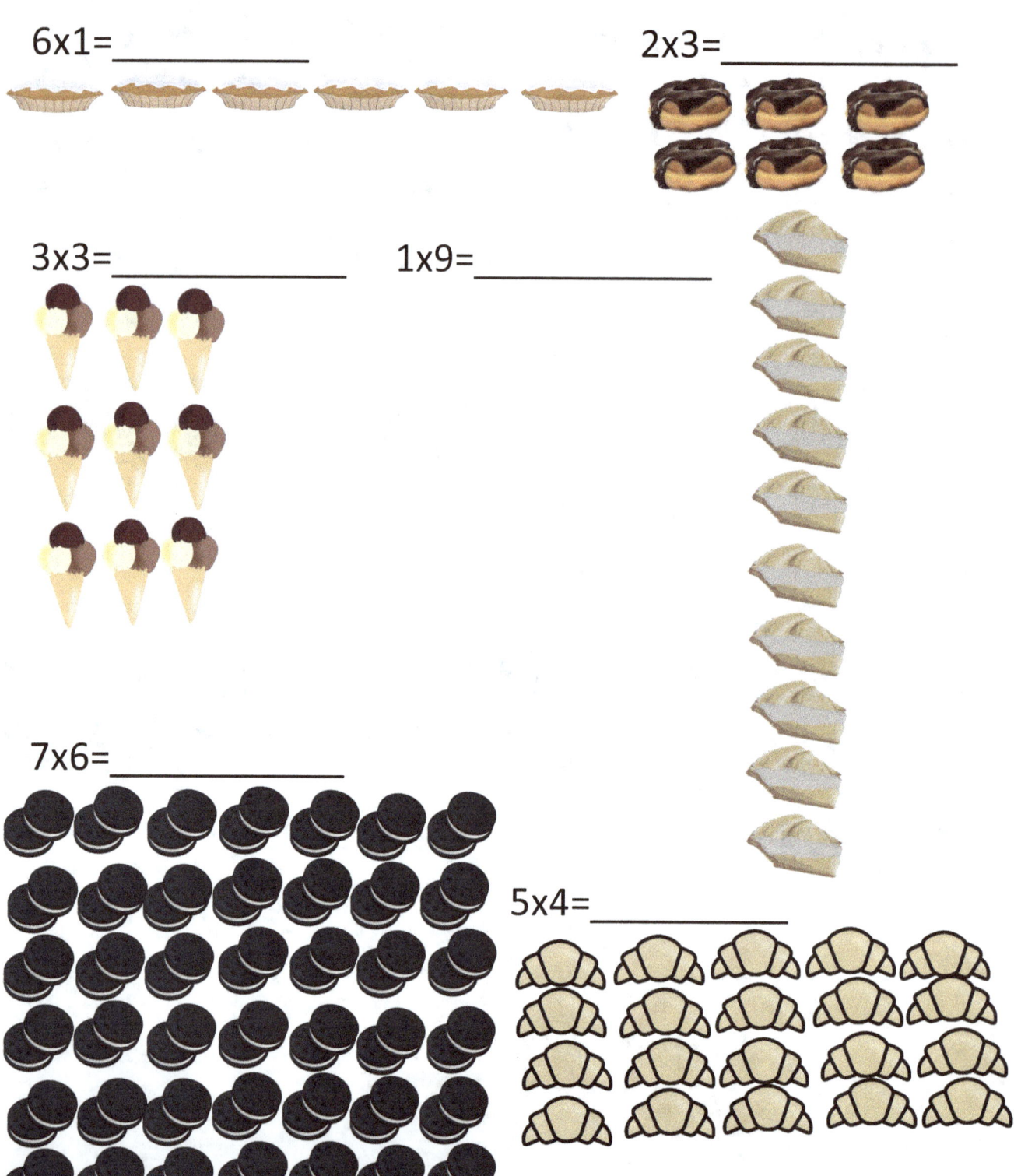

Name_____ Date_____

 Round the Order to the Nearest Hundred

Lesedi sometimes wants estimates of each order so she can prepare her supplies better, but today she needs your help.

Cakes- 582   924   418   163   347   298

_____   _____   _____   _____   _____   _____

Pies- 550   777   367   640   881   93

_____   _____   _____   _____   _____   _____

Brownies-145   253   985   533   768   333

_____   _____   _____   _____   _____   _____

Cookies-379   686   858   394   153   528

_____   _____   _____   _____   _____   _____

Fudge-136   542   707   324   658   366

_____   _____   _____   _____   _____   _____

Sourdough-538   309   858   452   625   896

_____   _____   _____   _____   _____   _____

Cupcakes-361   819   436   193   799   260

_____   _____   _____   _____   _____   _____

Dragon Lesedi's Magical Magnificent Bakery 2nd grade 1 Stelanie Marshall LLC

Name_____ Date_____

Sweet Tooth Orders. Count, chart and write the following.

_____   _____   _____   _____   _____

Orange   Purple   Red   Yellow   Blue

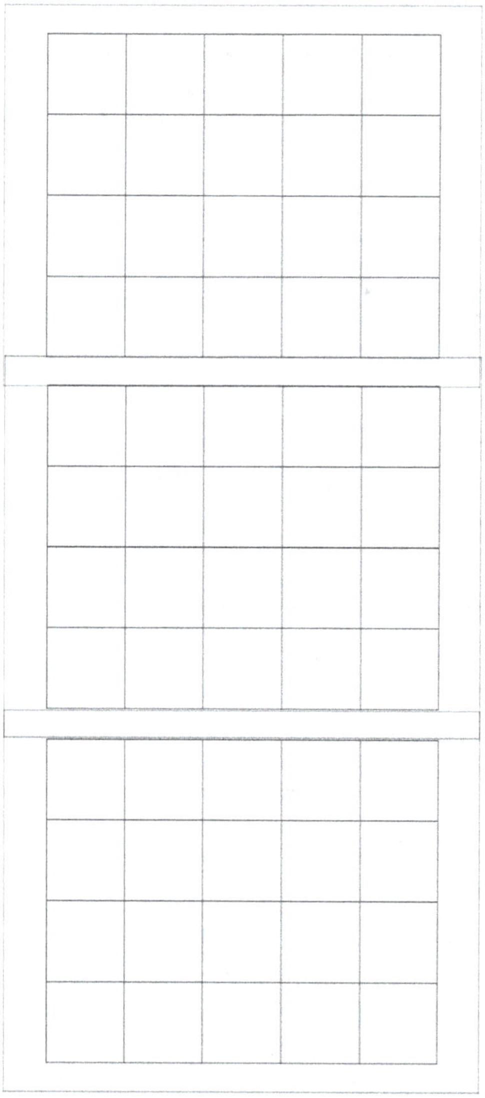

Name_____ Date_____

 Sweet Tooth Orders Questions.

1- Which color on the chart has the most?_____

2-Which color on the chart has the least? _____

3-Which color has 5 (five) orders?_____

4-Which has more, red or yellow?_____

5-Which has more orange or purple?_____

6-Which has the least, blue or purple?_____

7- Which has the least, orange or blue?_____

8-Which has the more, purple or red?_____

9-Which color has 6 orders?_____

10-Which color might run out the fastest?_____

Name_____ Date_____

One of the queens desires many treats from the bakery. She only wants the boxes of sweets that have ten in the box. Circle the ones that equal 10 green and the ones that do not orange.

5+4=_____   12-4=_____   6+4=_____   2+5=_____   30-20=_____

2x5=_____   11-2=_____   3x3=_____   1+9=_____   10x4=_____

0x10=_____   2+2+8=_____

5+5=_____   3+3+4=_____

2+5+1=_____   1x10=_____

2+8=_____   2x8=_____

5+5=_____   4+2+4=_____

9+0+1=_____   5x2=_____   3+2+5=_____   6x10=_____

8+3=_____   6+2+3=_____   8+1+1=_____   10x1=_____

5x5=_____   6x5=_____   6+5+8=_____   10x8=_____

Name_____ Date_____

Fresh out of the oven. Count, chart and write the following.

| _____ | _____ | _____ | _____ | _____ |
| Blue | Pink | Brown | Red | Purple |

Name_____ Date_____

Fresh Out of The Oven Questions.

1- Which color on the chart has the most?_____

2-Which color on the chart had the least? _____

3-Which color had 6 (six) orders?_____

4-Which has more, red or blue?_____

5-Which has more, brown or purple?_____

6-Which has the least, blue or purple?_____

7- Which has the least, pink or brown?_____

8-Which has more, brown or red?_____

9-Which color has 7 orders?_____

10-Which color might run out the slowest?_____

Name_____ Date_____

The two princesses always share but they only want desserts that come in even numbers. Circle even numbers pink and odd grey.

6+3+1=_____   5+5+5=_____   6+7=_____   4x5=_____

6+8+6=_____   9+4+4=_____   2x3=_____   3+5+7=_____

13+6=_____   5x5=_____   10x1=_____   8x2=_____

10+1=_____      6+6=_____

9+3=_____   7+7+3=_____

9x3=_____   10+15+5=_____

12x1=_____   3x0=_____   15+2=_____   12+1=_____

6x6=_____   2x7=_____   32+3=_____   7+2=_____

20+50=_____   3+3=_____   5x10=_____   3x3=_____

3x4=_____   2x6=_____   7x1=_____   24+1=_____

22+0=_____   4+3=_____   6+6=_____   6+2=_____

Name_____ Date_____

One of the prince's pets loves pancakes, bacon and sunny side eggs, but he only wants plates that have an odd number of pancakes. Circle odd numbers blue and even orange.

11x0=_____   2+3=_____   4+1+2=_____   3x3=_____

3x4=_____

4+4=_____

3+3=_____   3x2=_____

5+3+7=_____   11+0=_____

4x5=_____   5x1=____

5x5=_____   3+5=_____

4x4=_____   9x3=_____

7x0=_____   2+7=_____

5+4=_____   11x0=_____   12+1=_____   4+4=_____

7x2=_____   12x1=_____   5+5=_____   10+3+4=_____

1+8+5=_____   9x1=_____   0x9=_____   9+1=_____

Name_____ Date_____

Fred the fairy loves chocolate chip cookies, but with cookies with a prime number amount of chips. To help you out 1,2,3 5,7, 11,13,17,19,23,29 and 31 in this assignment. Circle prime yellow and non-prime purple.

3+9=_____

31-30=_____   29-6=_____

5x2=_____

7+2+2=_____

11x1=_____

4+15=_____

5x5=_____

14+14=_____

4+3=_____

10+19+2=_____

9+8=_____

7-2=_____   9-3=_____

1x7=_____

5x6=_____   10x7=_____

10+13=_____

7x0=_____   16+3=_____

3x3=_____

15+7+7=_____   6+5+1=_____

15+4=_____

28-23-3=_____   3x0=_____

6+5=_____   7x2=_____   6x7=_____   3x1=_____

Name_____ Date_____

Charlie the pirate loves gems with 5 (pentagons) and 10 (Decagons) sides. Circle the numbers divisible by 5 brown and not orange.

5+2+2=_____   6+3+7=_____   23-18=_____   30-5-10=_____

5+3=_____   2+5=_____

17-2=_____   12+5+13=_____

5x2= _____   18-4=_____   10+14+5=_____

29-4=_____   20+8+3=_____   5x3=_____   0x7=_____

10x2=_____   30-6-7=_____

5x3=_____   19+21=_____

7+6+3=_____   10x0=_____

3+16+11=_____   10+5=_____

10x5=_____   30-19-7=_____   5x5=_____   31+8=_____

5+5=_____   10x7=_____   8+10=_____   22+23=_____

Dragon Lesedi's Magical Magnificent Bakery 2nd grade 1 Stelanie Marshall LLC

Name_____ Date_____

This one might be a little tricky so take your time. This yellow knight enjoys cupcakes with sprinkles. Yes, with sprinkles but he wants them to have either less than or equal to 12 or 25 and greater sprinkles. Red if yes orange if no.

7+6=_____   6x5=_____   12x1=_____   14+10=_____

15+8=_____   14+5+5=_____   5x5=_____

12+12=_____   12+1=_____   15+15=_____

10x3=_____   14+14=_____   40+20=_____

13+13=_____   25-0=_____

12x2=_____   7+19=_____   6x4=_____

1x12=_____   4x4=_____

6x6=_____   5+5=_____

9x3=_____   2x10=_____   10+3=_____

7x6=_____   7x7=_____   5x5=_____

6+5=_____   5x3=_____   6+6=_____   5+3=_____

Name_____ Date_____

Shada loves big plates of pancakes. Either with 33 to 45 or 50 to 82 pancakes in her order. Circle the numbers she would order pink and ones she would refuse blue.

3x10=_____   6x6=_____   9+9=_____   90-30-8=_____

12x1=_____   11x3=_____   4+11=_____

10+5=_____   40+40+2=_____   7x7=_____

8x6=_____

6+7=_____   11x7=

_____

5x9=_____

93-31-14=_____   5x6=_____

7x8=_____   4x8=_____

9x10=_____   11+22=_____

70-35=_____   30+20=_____   60-29=_____   33+33=_____

9x9=_____   12x3=_____   7+16+7=_____   80-5-30=_____

10x5=_____   17+18+13=_____   54-45=_____   10x6=_____

Dragon Lesedi's Magical Magnificent Bakery 2nd grade 1 Stelanie Marshall LLC

Name_____ Date_____

Who's Who Ticket Number? Color in the correct square.

	293	292	308	300

1 The blue knight number has a 3 in the hundreds place.

2 The unicorn has an even number

3 The dragon's number is divisible by 10

Name_____ Date_____

Recipe Math. Grumpy Sloppy Ogre Cake.

1. Turn the oven to 639-289=_____°

In a bowl mix-

2. 23-9=_____ cups of . 3. 97-79=_____ cups of

4. 2x3=_____ cups of 5. Sift 3x4=_____ tsp of

6. 5x1=_____ tsp of  . 7. 4x2=_____ tsp of

8. 28-23=_____ sticks of melted  . 9. 5x2=_____

10. 62-54=_____ tsp of  .11. Put 6x2=_____ tsp of

12. Place in the  for 7x5=_____ minutes.

Frosting

13. Pour 6x1 of  in a bowl. 14 Mix 4x3=_____ tsp of .

15. 2x7=_____ tbs of  . 16. Take cake out and let it cool for 6x5=_____ minutes.

17. Will serve up to 4x10 =_____ people or 18. 8x3=_____ ogres.

Name_____ Date_____

Who's Who Ticket Number? Color in the correct square.

	503	495	459	508

1. Blue warrior's ticket has a 5 that is not in the hundreds place.

2. Yellow knight's ticket is divisible by 5.

3. Red rose fairy is an odd number.

Dragon Lesedi's Magical Magnificent Bakery 2nd grade 1 Stelanie Marshall LLC

Name_____ Date_____

Dessert to What Ticket Number? Color in the correct square.

562				
910				
806				
695				

1 The slice of cake goes with ticket number that is divisible by 5.

2 806 ticket asks for "extra sprinkles".

3 The doughnut goes with ticket number with the 9 that is not in the tens place.

Name_____ Date_____

Subtraction With a Pinch of Borrowing.

| 1) 183
- 92 | 2) 767
-399 | 3) 598
-239 | 4) 664
-381 | 5) 441
-150 |

| 6) 885
-548 | 7) 315
- 63 | 8) 609
-413 | 9) 723
-191 | 10) 995
-647 |

| 11) 723
-673 | 12) 930
-124 | 13) 446
-255 | 14) 890
-325 | 15) 534
-371 |

| 16) 943
-490 | 17) 958
- 74 | 18) 735
-507 | 19) 642
-271 | 20) 598
-299 |

Multiplication-

| 1
X 7 | 9
x3 | 6
x4 | 8
x2 |

Dragon Lesedi's Magical Magnificent Bakery 2nd grade 1 Stelanie Marshall LLC

Name_____ Date_____

Adding Up the Orders with Multiplication.

10x2=_____

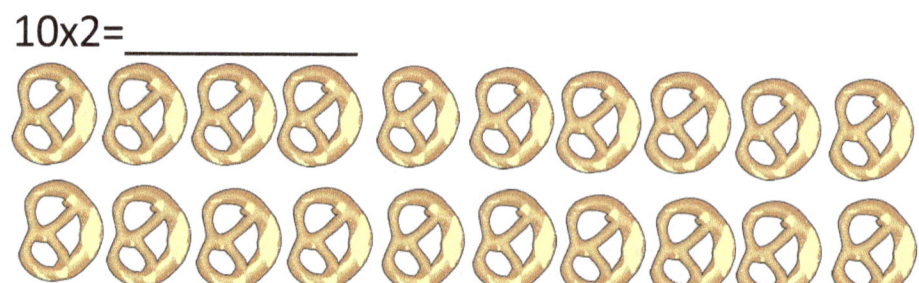

3x6=_____   5x2=_____

Draw the following:

A) Blue doughnuts 7x_____=28   B) Brown bacon 3x_____=15

Name_____ Date_____

Help with the orders. Addition, Subtraction and Calendar.

July						
Sunday	Monday	Tuesday	Wednesday	Thursday	Friday	Saturday
			1	2	3	4
5	6	7	8	9	10	11
12	13	14	15	16	17	18
19	20	21	22	23	24	25
26	27	28	29	30	31	

1) $156+253=$_____ and to be made for the festival on the 18th.
What weekday is that?_____

2) $364+193=$_____ for the king's huge royal monthly theater party on the 3rd third Wednesday of the month. What is the date?_____

3) $95+446=$_____ for the dragon's fire event on the 7/21/xxxx. What weekday is that on?

4) $630+86=$_____ for the fisherman's competition for the 17th. What weekday is that on?

_____

Dragon Lesedi's Magical Magnificent Bakery 2nd grade 1 Stelanie Marshall LLC

Name_____ Date_____

Help with the orders. Addition, Subtraction and Calendar. Page 2

5) 94+67=_____ to be read for the last Wednesday of the month. What is the date of that day?

_____

6) 578+341=_____ and 692+409=_____ to be ready for the 30th for the fairy's tea and breakfast. What weekday is this event on?

_____

7) 334+295=_____ for the prince's event on the 11th but need to prepare all the supplies two days before. What is the date that the supplies are needed and what day is the event?

_____   _____

8) 472+85=_____ to be made for the fairy godmother graduation practice on 7/9/xxxx. What weekday is that on?

_____

9) 712+493=_____ slices of for the last Sunday for the royal picnic. What is the date?

_____

10) The warlock ordered 8x8=_____ for his biweekly magic bonfire on the 1st and 3rd Saturday. What are the two dates?

_____   _____

Name_____ Date_____

Work Area/ Chefs Corner

Name_____ Date_____

Answers

Pg3- 1. 558 2. 8 3. 80 4. 15 5. 345 6. 30 7. 31 8. 4 9 < 10 > 11> 12<
13- 20,50,70, 100 14- 20, 80, 100 15- 10,15,35,60,70,75,85 16- .81
17- .70 18- .48

Pg4- 1-648 2-927 3-893 4-1,353 5-491 6-843 7-915 8-1,152 9-859 10-994 11-142
12-1,370

Pg5- 1< 2> 3< 4> 5< 6< 7> 8< 9< 10< 11< 12< 13< 14< 15> 16> 17<

Pg6- 1-9 2-11 3-4 4-1 5-17 6-4 7-10 8-8 9-21 10-8,22,76,110,134
11-100,65,20,15,5 12-7,77,109,29,51,139 13-80,170,140,190 14-37

Pg7- 1) 300 90 1 2) 100 0 7 3) 500 40 5 4) 900 30 4 5) 700 20 9 6) 800 60 3 7) 400
30 2

Pg8- 1) 900 50 6 2) 300 40 5 3) 700 10 8 4) 200 30 9 5) 800 60 7 6) 100 80 0 7) 400
20 1

Pg9- 1) 7000 900 80 5 2) 3000 400 70 1 3) 5000 300 40 9 4) 1000 200 30 7 5) 4000
600 0 3 6) 2000 700 50 8 7) 7000 100 90 5 8) 6000 500 10 2

pg10- 5000 300 30 2 2) 1000 800 90 1 3) 7000 100 30 9 4) 6000 500 70 4 5) 3000
200 90 7 6) 4000 700 0 3 7) 9000 400 10 5 8) 8000 900 20 6

pg11- Queen 836 Dragon 520 Blue Knight 559 Rose Fairy 605

Pg12- Sheep 288 Male fairy 294 Pirate 265 Yellow Knight 379

Pg13- Row1 251 509 215 Row2 97 107 499 Row3 680 65 193 Row4 138 171 199

Pg14- Row <>> Row 2 ><= Row 3 >< Row 4 <> Row 5 <= Row 6 << Row 7 <>>

Name_____ Date_____

Pg15- 1/2/2xxx January 2, 2/1/2000 February 1, 11/6/2xxx November 6, 10/7/2xxx October 7, 12/7/2xxx December 7, 6/11/2xxx June 11, 7/12/2xxx July 12.

Pg16-17 1) 2x6 12 3x5 15 2)5x8 40 6x8 48 3)10x10 100 8x7 56 4) 3x7 21 4x9 36 5) 8x8 64 5x10 50 6) 2x3 6 4x4 16 7) 8x1 3x2 6 8) 5x9 45 8x3 24 9) 9x5 45 11x7 77 10)5x4 20 7x8 56 11) 6x7 42 3x10 30 12) 3x2 6 8x6 48

Pg19 1-20 2-12 3-28 4-25 5-64 6-1 **Pg20** 66 27 4 16 6 **Pg21** 18 35 24 30 24 **Pg22** 6 6 9 9 42 20

Pg23- Cakes 600 900 400 200 300 300 Pies 600 800 400 600 900 100 Brownies 100 300 1000 500 800 300 Cookies 400 700 900 400 200 500 Fudge 100 500 700 300 700 400 Sourdough 500 300 900 500 600 900 Cupcakes 400 800 400 200 800 300

Pg24-25 Orange 5 Purple 4 Red 7 Yellow 10 Blue 6 1-Yellow 2- Purple 3- Orange 4-Yellow 5-Orange 6-Purple 7-Orange 8-Red 9-Blue 10-Yellow

Pg26 5+4=_9o_   12-4=_8o_   6+4=_10g_   2+5=_7o_   30-20=_10g_

2x5=_10g_ 11-2=_9o_   3x3=_9o_ 1+9=_10g_   10x4=_40o_

0x10=_0o_   2+2+8=_12o_

5+5=_10g_   3+3+4=_10g_

2+5+1=_8o_   1x10=_10g_

2+8=_10g_   2x8=_16o_

Name_____ Date_____

5+5=___10g___   4+2+4=___10g___

9+0+1=___10g___   5x2=___10g___   3+2+5=___10g___   6x10=___60o___

8+3=___11o___   6+2+3=___11o___   8+1+1=___10g___   10x1=___10g___

5x5=___25o___   6x5=___30o___   6+5+8=___19o___   10x8=___80o___

Pg27-28 Blue 6 Pink 7 Brown 9 Red 5 Purple 12 1 Purple 2 Red 3 Blue 4 Blue 5 Purple 6 Blue 7 Pink 8 Brown 9 Pink 10 Red

Pg29

6+3+1=___10p___   5+5+5=___15g___   6+7=___13g___   4x5=___20p___

6+8+6=___20p___   9+4+4=___17g___   2x3=___6p___   3+5+7=___15g___

13+6=___19g___   5x5=___25g___   10x1=___10p___   8x2=___16p___

10+1=___11g___   6+6=___12p___

9+3=___12p___   7+7+3=___17g___

9x3=___27g___   10+15+5=___30p___

Dragon Lesedi's Magical Magnificent Bakery 2nd grade 1 Stelanie Marshall LLC

Name_____ Date_____

12x1=___12p___   3x0=___0p___   15+2=___17g___   12+1=___13g___

6x6=___36p___   2x7=___14p___   32+3=___35g___   7+2=___9g___

20+50=__70p___   3+3=___6p___   5x10=__50p___   3x3=___9g___

3x4=___12p___   2x6=___12p___   7x1=__7g___   24+1=___25g___

22+0=_22p___   4+3=_7g___   6+6=___12p___   6+2=__8p___

Pg30

11x0=___0o___   2+3=___5b___   4+1+2=__7b___   3x3=___9b___

3x4=___12o___   4+4=__8o___

3+3=__6o___   3x2=___6o___

5+3+7=_15b___   11+0=_11b___

4x5=_20o___   5x1=_5b___

5x5=___25b___   3+5=__8o___

4x4=_16o___   9x3=___27b___

Name_____ Date_____

7x0=____0o____   2+7=____14o____

5+4=____9b____   11x0=__0o____   12+1=____13b____   4+4=____8o____

7x2=____14o____   12x1=__12o____   5+5=____10o____   10+3+4=____17b____

1+8+5=____14o____   9x1=____9b____   0x9=____0o____   9+1=____10o____

Pg31

3+9=____12p____   31-30=____61y____   29-6=____23y____

5x2=_10p____   7+2+2=__11y____

11x1=_11y____   4+15=____19y____

5x5=____10p____   14+14=____28p____

4+3=____7y____   10+19+2=____31y____

9+8=____17y____   7-2=____5y____   9-3=____6p____

1x7=____7y____   5x6=__30p____   10x7=__70p____

Name_____ Date_____

10+13=___23y___

7x0=___0___   16+3=__19y____

3x3=___9y___

15+7+7=___29y___   6+5+1=__12p___

15+4=___19y___

28-23-3=___2y___   3x0=___0___

6+5=___11y___   7x2=___14p___   6x7=___42p___   3x1=___3y___

Pg32

5+2+2=___9o___   6+3+7=___16o___   23-18=___5b___   30-5-10=_15b___

5+3=___8o___

2+5=___7o___

17-2=___15b___

12+5+13=___30b___

5x2=___10b___   18-4=__14o__

10+14+5=___29o___

29-4=_25b___   20+8+3=___31o___  5x3=_15b___   0x7=__0o___

10x2=___20b___

30-6-7=___17o___

5x3=___15b___

19+21=___40b___

7+6+3=___16o___

10x0=___0o___

Name_____ Date_____

3+16+11=___30b___   10+5=___15b___

10x5=___50b___   30-19-7=___4o___   5x5=__25b__   31+8=___39o___

5+5=___10b___   10x7=___70b___   8+10=___18o___   22+23=___45o___

Pg33

7+6=___13o___   6x5=___30r___   12x1=__12r__   14+10=___25r___

15+8=___23o___   14+5+5=___19o___   5x5=__25r__

12+12=__24o___   12+1=___13o___   15+15=__30r__

10x3=___30r___   14+14=___24o___   40+20=__60r__

13+13=___26r___   25-0=___25r___

12x2=__24o___   7+19=__26r__   6x4=___24o___

1x12=   4x4=__16o__

___12r___

6x6=   5+5=___10r___

___36r___

Name_____ Date_____

9x3=

_____27r_____

7x6=____42r_____

2x10=____20o_____   10+3=____13o_____

7x7=_____49r_____   5x5=_____25r_____

6+5=____11o____   5x3=____15o____   6+6=___12r____   5+3=____8r____

Pg34

3x10=___30b_____   6x6=____36p____   9+9=___18b____   90-30-8=_52p____

12x1=____12b____   11x3=____33p____   4+11=____15b____

10+5=___15b___   40+40+2=__82p_____   7x7=__49b___

8x6=___48b___

6+7=____13b____   11x7=

_____77p____

5x9=_45p____

93-31-14=____48b____   5x6=_30b____

7x8=__56p____   4x8=___32b____

9x10=___90b____   11+22=_33p__

70-35=_35p____   30+20=_50p___   60-29=__31b____   33+33=___66p____

Dragon Lesedi's Magical Magnificent Bakery 2nd grade 1 Stelanie Marshall LLC

Name_____ Date_____

9x9=___81p___   12x3=___36p___   7+16+7=___30b___   80-5-30=___45p___

10x5=___50p___   17+18+13=___48b___   54-45=___9b___   10x6=___60p___

Pg35 Blue knight 308 Green Monster 293 Dragon 300 Unicorn 292

Pg36 1-350 2-14 3-18 4-6 5-12 6-5 7-8 8-5 9-10 10-8 11-12 12-35 13-6 14-12 15-14 16-30 17-40 18-24

Pg37 yellow knight 495 green fairy 508 blue warrior 459 rose fairy 503

Pg38 slice of cake 695 doughnut 910 cupcake 806 wedding cake 562

Pg39 1)91 2)368 3)359 4)283 5)291 6)337 7)252 8)196 9)532 10)348 11)50 12)806 13)191 14)565 15)163 16)453 17)884 18)228 19)371 20)299 Mul 7,27,24,16

Pg40 20,18,10,4,5 Pg41

Pg41-42 1)409,Saturday 2)557,7/15/2xxx 3) 541,Tues 4)716/Friday 5)161, 7/29/2xxx 6) 919,1,101,Thursday 7)629 7/9/2xxx, Saturday 8) 557, Thursday 9-1,205, 7/26/2xxx 10)64 7/4/ 2xxx, 7/18/2xxx

Name_____ Date_____

1	2	3	4	5	6	7	8	9	10	11	12	13	14
15	16	17	18	19	20	21	22	23	24				
25	26	27	28	29	30	31	32	33	34				
35	36	37	38	39	40	41	42	43	44				
45	46	47	48	49	50	51	52	53	54				
55	56	57	58	59	60	61	62	63	64				
65	66	67	68	69	70	71	72	73	74				
75	76	77	78	79	80	81	82	83	84				
85	86	87	88	89	90	91	92	93	94				
95	96	97	98	99	100								

Name_____ Date_____

Name_____ Date_____

Books to Look Out For

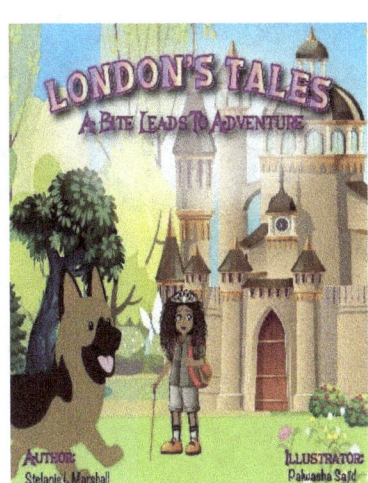

Dragon Lesedi's Magical Magnificent Math Bakery series

Many other workbooks for different grades.

More children's books are always in the process.

Journals, notebooks, stickers, and other items to be up by September 2021.

Check for updates on https://www.facebook.comAccessoriesinfinitymi

All by Stelanie Marshall LLC

Name_____ Date_____

MULTIPLICATION

$$
\begin{array}{r} 0 \\ \times\,1 \\ \hline \end{array}
\qquad
\begin{array}{r} 0 \\ \times 2 \\ \hline \end{array}
\qquad
\begin{array}{r} 0 \\ \times 3 \\ \hline \end{array}
\qquad
\begin{array}{r} 0 \\ \times 4 \\ \hline \end{array}
\qquad
\begin{array}{r} 0 \\ \times 5 \\ \hline \end{array}
$$

$$
\begin{array}{r} 0 \\ \times 6 \\ \hline \end{array}
\qquad
\begin{array}{r} 0 \\ \times 7 \\ \hline \end{array}
\qquad
\begin{array}{r} 0 \\ \times 8 \\ \hline \end{array}
\qquad
\begin{array}{r} 0 \\ \times 9 \\ \hline \end{array}
\qquad
\begin{array}{r} 0 \\ \times 10 \\ \hline \end{array}
$$

$$
\begin{array}{r} 0 \\ \times 11 \\ \hline \end{array}
\qquad
\begin{array}{r} 0 \\ \times 12 \\ \hline \end{array}
$$

Name_____ Date_____

$$\begin{array}{r} 1 \\ \times\,1 \\ \hline \end{array}$$ $$\begin{array}{r} 1 \\ \times\,2 \\ \hline \end{array}$$ $$\begin{array}{r} 1 \\ \times\,3 \\ \hline \end{array}$$ $$\begin{array}{r} 1 \\ \times\,4 \\ \hline \end{array}$$ $$\begin{array}{r} 1 \\ \times\,5 \\ \hline \end{array}$$

$$\begin{array}{r} 1 \\ \times\,6 \\ \hline \end{array}$$ $$\begin{array}{r} 1 \\ \times\,7 \\ \hline \end{array}$$ $$\begin{array}{r} 1 \\ \times\,8 \\ \hline \end{array}$$ $$\begin{array}{r} 1 \\ \times\,9 \\ \hline \end{array}$$ $$\begin{array}{r} 1 \\ \times\,10 \\ \hline \end{array}$$

$$\begin{array}{r} 1 \\ \times\,11 \\ \hline \end{array}$$ $$\begin{array}{r} 1 \\ \times\,12 \\ \hline \end{array}$$

Name_____ Date_____

$$\begin{array}{r} 2 \\ \times\ 1 \\ \hline \end{array} \quad \begin{array}{r} 2 \\ \times 2 \\ \hline \end{array} \quad \begin{array}{r} 2 \\ \times 3 \\ \hline \end{array} \quad \begin{array}{r} 2 \\ \times 4 \\ \hline \end{array} \quad \begin{array}{r} 2 \\ \times 5 \\ \hline \end{array}$$

$$\begin{array}{r} 2 \\ \times 6 \\ \hline \end{array} \quad \begin{array}{r} 2 \\ \times 7 \\ \hline \end{array} \quad \begin{array}{r} 2 \\ \times 8 \\ \hline \end{array} \quad \begin{array}{r} 2 \\ \times 9 \\ \hline \end{array} \quad \begin{array}{r} 2 \\ \times 10 \\ \hline \end{array}$$

$$\begin{array}{r} 2 \\ \times 11 \\ \hline \end{array} \quad \begin{array}{r} 2 \\ \times 12 \\ \hline \end{array}$$

Name_____ Date_____

$$3 \times 1$$ $$3 \times 2$$ $$3 \times 3$$ $$3 \times 4$$ $$3 \times 5$$

$$3 \times 6$$ $$3 \times 7$$ $$3 \times 8$$ $$3 \times 9$$ $$3 \times 10$$

$$3 \times 11$$ $$3 \times 12$$

Name_____ Date_____

$$
\begin{array}{r} 4 \\ \times\ 1 \\ \hline \end{array}
\qquad
\begin{array}{r} 4 \\ \times 2 \\ \hline \end{array}
\qquad
\begin{array}{r} 4 \\ \times 3 \\ \hline \end{array}
\qquad
\begin{array}{r} 4 \\ \times 4 \\ \hline \end{array}
\qquad
\begin{array}{r} 4 \\ \times 5 \\ \hline \end{array}
$$

$$
\begin{array}{r} 4 \\ \times 6 \\ \hline \end{array}
\qquad
\begin{array}{r} 4 \\ \times 7 \\ \hline \end{array}
\qquad
\begin{array}{r} 4 \\ \times 8 \\ \hline \end{array}
\qquad
\begin{array}{r} 4 \\ \times 9 \\ \hline \end{array}
\qquad
\begin{array}{r} 4 \\ \times 10 \\ \hline \end{array}
$$

$$
\begin{array}{r} 4 \\ \times 11 \\ \hline \end{array}
\qquad
\begin{array}{r} 4 \\ \times 12 \\ \hline \end{array}
$$

Name_____ Date_____

$$5 \times 1$$ $$5 \times 2$$ $$5 \times 3$$ $$5 \times 4$$ $$5 \times 5$$

$$5 \times 6$$ $$5 \times 7$$ $$5 \times 8$$ $$5 \times 9$$ $$5 \times 10$$

$$5 \times 11$$ $$5 \times 12$$

Name_____ Date_____

$$6 \times 1 \qquad 6 \times 2 \qquad 6 \times 3 \qquad 6 \times 4 \qquad 6 \times 5$$

$$6 \times 6 \qquad 6 \times 7 \qquad 6 \times 8 \qquad 6 \times 9 \qquad 6 \times 10$$

$$6 \times 11 \qquad 6 \times 12$$

Name_____ Date_____

$$7 \times 1$$ $$7 \times 2$$ $$7 \times 3$$ $$7 \times 4$$ $$7 \times 5$$

$$7 \times 6$$ $$7 \times 7$$ $$7 \times 8$$ $$7 \times 9$$ $$7 \times 10$$

$$7 \times 11$$ $$7 \times 12$$

Name_____ Date_____

$$\begin{array}{r} 8 \\ \times\ 1 \\ \hline \end{array} \quad \begin{array}{r} 8 \\ \times 2 \\ \hline \end{array} \quad \begin{array}{r} 8 \\ \times 3 \\ \hline \end{array} \quad \begin{array}{r} 8 \\ \times 4 \\ \hline \end{array} \quad \begin{array}{r} 8 \\ \times 5 \\ \hline \end{array}$$

$$\begin{array}{r} 8 \\ \times 6 \\ \hline \end{array} \quad \begin{array}{r} 8 \\ \times 7 \\ \hline \end{array} \quad \begin{array}{r} 8 \\ \times 8 \\ \hline \end{array} \quad \begin{array}{r} 8 \\ \times 9 \\ \hline \end{array} \quad \begin{array}{r} 8 \\ \times 10 \\ \hline \end{array}$$

$$\begin{array}{r} 8 \\ \times 11 \\ \hline \end{array} \quad \begin{array}{r} 8 \\ \times 12 \\ \hline \end{array}$$

Name_____ Date_____

$$9 \times 1 \qquad 9 \times 2 \qquad 9 \times 3 \qquad 9 \times 4 \qquad 9 \times 5$$

$$9 \times 6 \qquad 9 \times 7 \qquad 9 \times 8 \qquad 9 \times 9 \qquad 9 \times 10$$

$$9 \times 11 \qquad 9 \times 12$$

Name_____ Date_____

$$10 \times 1 \qquad 10 \times 2 \qquad 10 \times 3 \qquad 10 \times 4 \qquad 10 \times 5$$

$$10 \times 6 \qquad 10 \times 7 \qquad 10 \times 8 \qquad 10 \times 9 \qquad 10 \times 10$$

$$10 \times 11 \qquad 10 \times 12$$

Name_____ Date_____

$$\begin{array}{r} 11 \\ \times\ 1 \\ \hline \end{array}$$
$$\begin{array}{r} 11 \\ \times 2 \\ \hline \end{array}$$
$$\begin{array}{r} 11 \\ \times 3 \\ \hline \end{array}$$
$$\begin{array}{r} 11 \\ \times 4 \\ \hline \end{array}$$
$$\begin{array}{r} 11 \\ \times 5 \\ \hline \end{array}$$

$$\begin{array}{r} 11 \\ \times 6 \\ \hline \end{array}$$
$$\begin{array}{r} 11 \\ \times 7 \\ \hline \end{array}$$
$$\begin{array}{r} 11 \\ \times 8 \\ \hline \end{array}$$
$$\begin{array}{r} 11 \\ \times 9 \\ \hline \end{array}$$
$$\begin{array}{r} 11 \\ \times 10 \\ \hline \end{array}$$

$$\begin{array}{r} 11 \\ \times 11 \\ \hline \end{array}$$
$$\begin{array}{r} 11 \\ \times 12 \\ \hline \end{array}$$

Name_____ Date_____

$$12$$
$$\underline{\times 1}$$
$$12$$
$$\underline{\times 2}$$
$$12$$
$$\underline{\times 3}$$
$$12$$
$$\underline{\times 4}$$
$$12$$
$$\underline{\times 5}$$

$$12$$
$$\underline{\times 6}$$
$$12$$
$$\underline{\times 7}$$
$$12$$
$$\underline{\times 8}$$
$$12$$
$$\underline{\times 9}$$
$$12$$
$$\underline{\times 10}$$

$$12$$
$$\underline{\times 11}$$
$$12$$
$$\underline{\times 12}$$